CRYSTAL LAKE PUBLIC LIBRARY
W9-AVM-047

Tick

Monica Harris

Heinemann Library
Chicago, Illinois

Customer Service 888-454-2279
Visit our website at www.heinemannlibrary.com

Designed by Ginkgo Creative, Inc.
Printed and bound in the United States by Lake Book Manufacturing, Inc.
Photo research by Scott Braut

07 06 05 04 03
10 9 8 7 6 5 4 3 2 1

Library of Congress Cataloging-in-Publication Data

Harris, Monica, 1964-
Tick / Monica Harris.
v. cm. — (Bug books)
Includes bibliographical references (p.).
Contents: What are ticks? — What do ticks look like? — How big are ticks? — How are ticks born? — How do ticks grow? — What do ticks eat? — What animals attack ticks? — Where do ticks live? — How do ticks move? — How long do ticks live? — What do ticks do? — How are ticks special? — Thinking about ticks.
ISBN: 1-40340-765-7 (HC), 1-40340-996-X (Pbk.)
1. Ticks—Juvenile literature. [1. Ticks.] I. Title. II. Series.
QL458 .H37 2003
595.4'29—dc21

2002004017

Acknowledgments

The author and publishers are grateful to the following for permission to reproduce copyright material: p. 4 Center for Disease Control; p. 5 Joe McDonald/Corbis; pp. 6, 11, 19 Stephen McDaniel; pp. 7, 12 Dr. James L. Castner; p. 8 Larry Mulvehill/Photo Researchers, Inc.; pp. 9, 17, 18 James H. Robinson; p. 10 R. Calentine/Visuals Unlimited; p. 13 Kenneth H. Thomas/Photo Researchers, Inc.; pp. 14, 21 Eye of Science/Photo Researchers; p. 15 Ken Eward/Science Source; p. 16 Joe McDonald/Visuals Unlimited; p. 20 Charles W. Mann; p. 22 Dan Suzio; p. 23 Jim Cummins/Taxi/Getty Images; p. 24 Anthony Bannister/Gallo Images/Corbis; p. 25 Kenneth Greer/Visuals Unlimited; p. 26 Susanne Danegger/NHPA; p. 27 David M. Phillips/Photo Researchers, Inc.; p. 28 Jim Occi; p. 29 Darwin Dale/Photo Researchers, Inc.

Illustration, p. 30, by Will Hobbs.
Cover photograph by Dr. James L. Castner.

Every effort has been made to contact copyright holders of any material reproduced in this book. Any omissions will be rectified in subsequent printings if notice is given to the publisher.

Special thanks to Dr.William Shear, Department of Biology, Hampden-Sydney College, for his review of this book.

Some words are shown in bold, **like this**. You can find out what they mean by looking in the glossary.

Contents

What Are Ticks?

Ticks are small **arachnids.** They are in the same group as spiders and scorpions. There are about 850 different kinds of ticks.

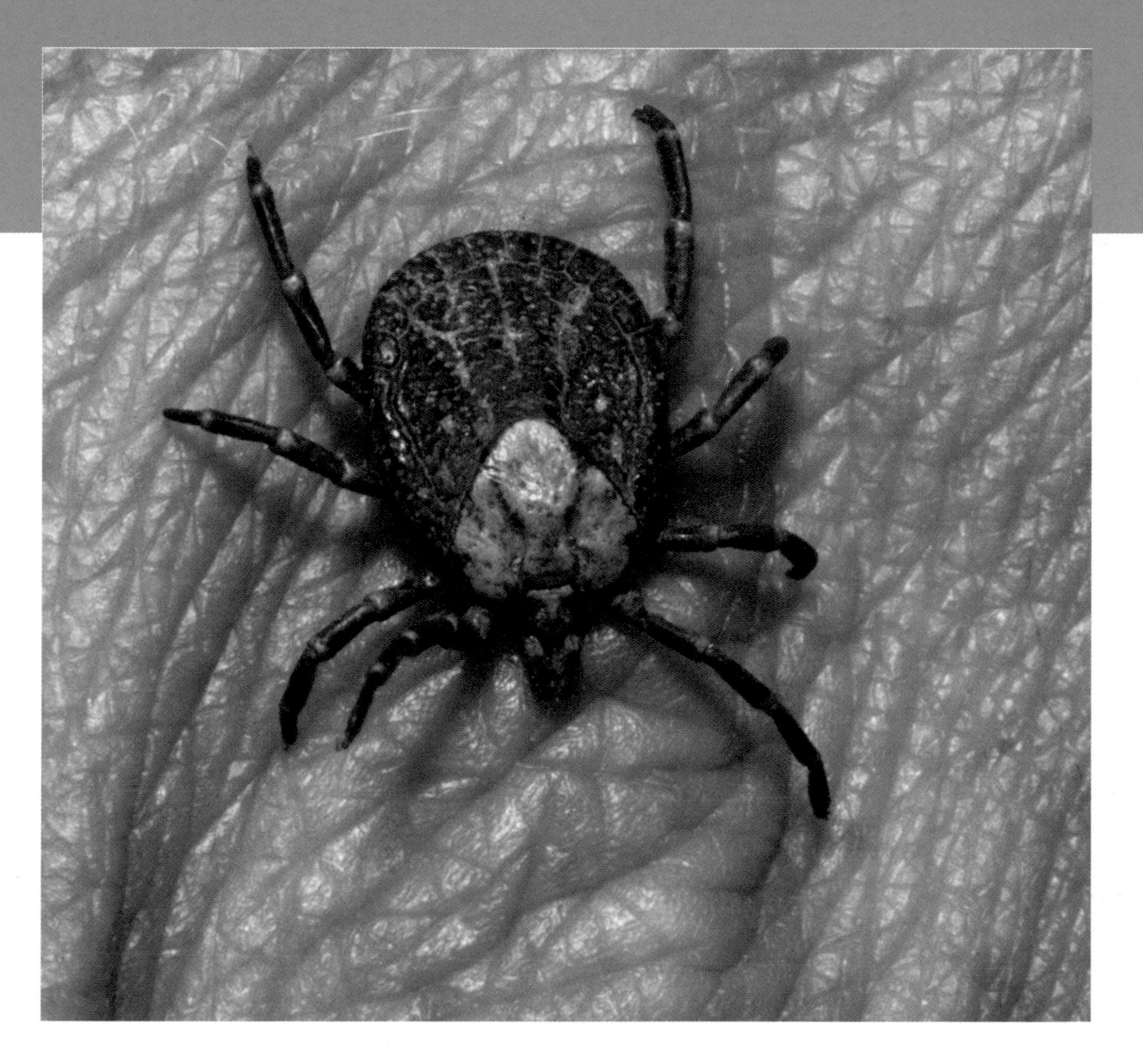

Ticks are **parasites.** They drink the blood of other animals. These animals are called **hosts.** Some ticks carry sicknesses like **Lyme disease.**

What Do Ticks Look Like?

Ticks are shaped like a teardrop. Most ticks are black or brown. Some are red, green, or yellow. They have small heads with special **mouthparts.** Most ticks do not have eyes.

Chicken ticks have soft bodies. Other ticks, like dog ticks, have a hard plate on their backs. It is called a **scutum.** Hard ticks spread **Lyme disease.**

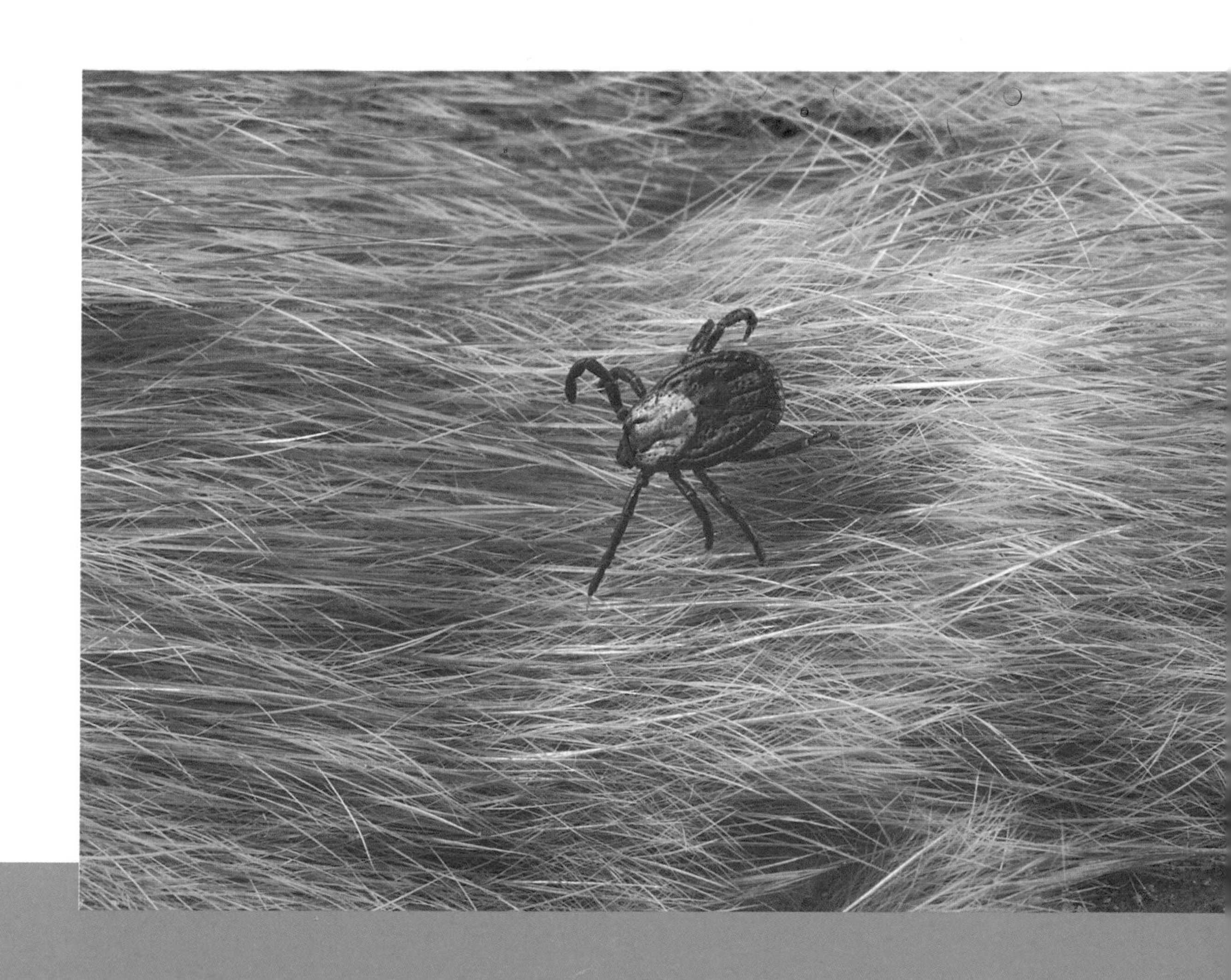

How Big Are Ticks?

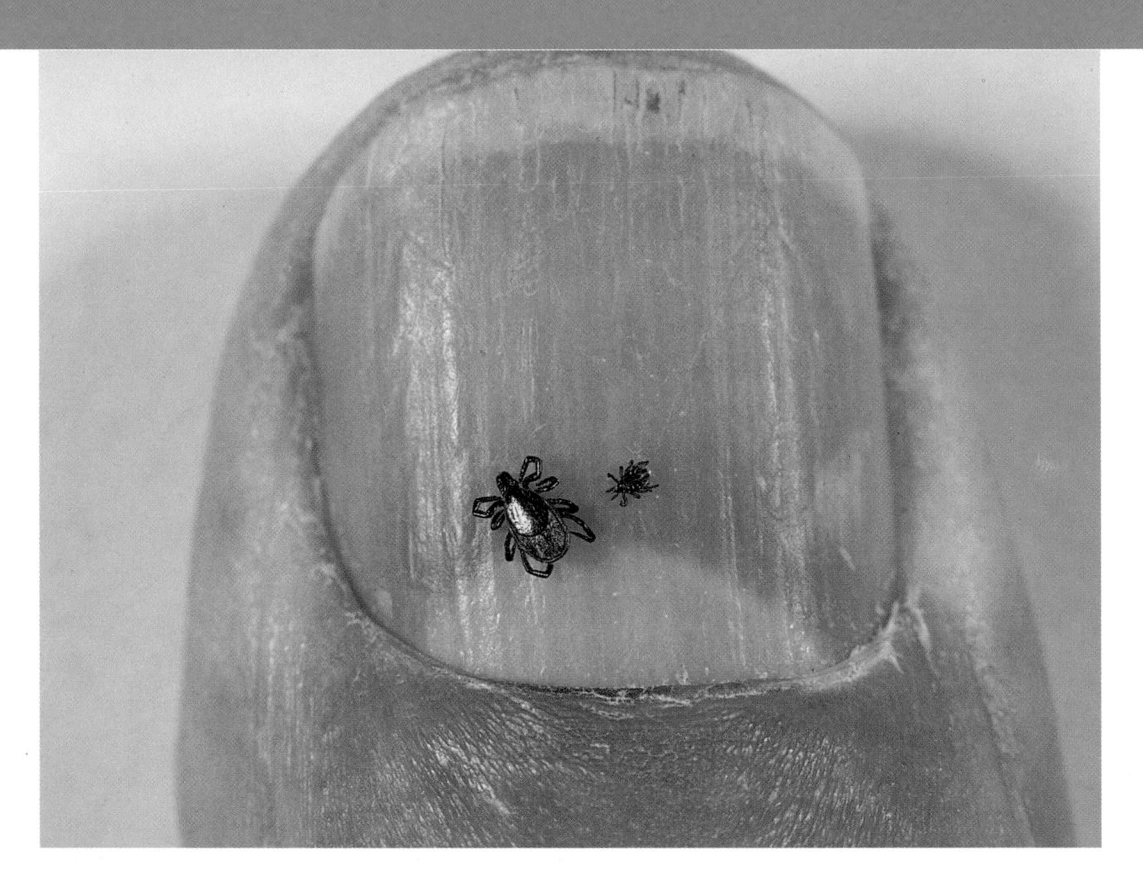

Young ticks are about the size of a pinhead. **Adult** ticks can be as small as a freckle or as big as an apple seed. **Females** are bigger than **males.**

Ticks get bigger when they drink blood. Their bodies can get 20 to 50 times bigger. This makes a feeding tick easier to see.

How Are Ticks Born?

Female ticks lay thousands of eggs. The eggs **hatch** in the spring. The little ticks are called **larvae.** They are light colored and have six legs.

Ticks have special **organs** in their legs. They let the tick know an animal is near. The tick lifts its front legs to grab the animal. This is called **questing.**

How Do Ticks Grow?

Larvae drink the **host's** blood for two to four days. Then they let go and fall to the ground. Larvae shed their skins for bigger ones. This is called **molting.**

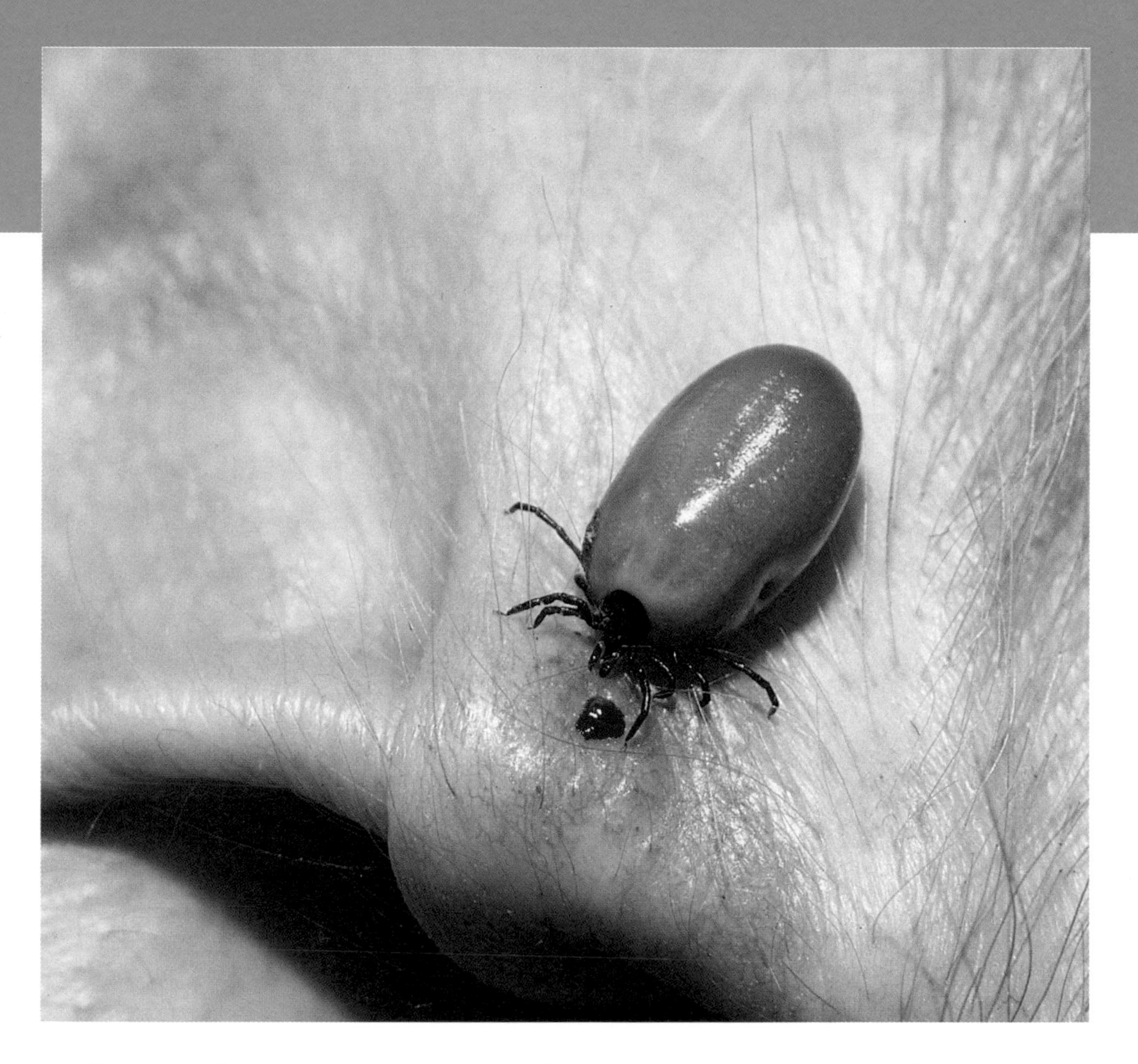

After they molt, ticks have eight legs and are called **nymphs.** Nymphs find one or two more hosts and drink their blood. They molt one last time to become **adults.**

What Do Ticks Eat?

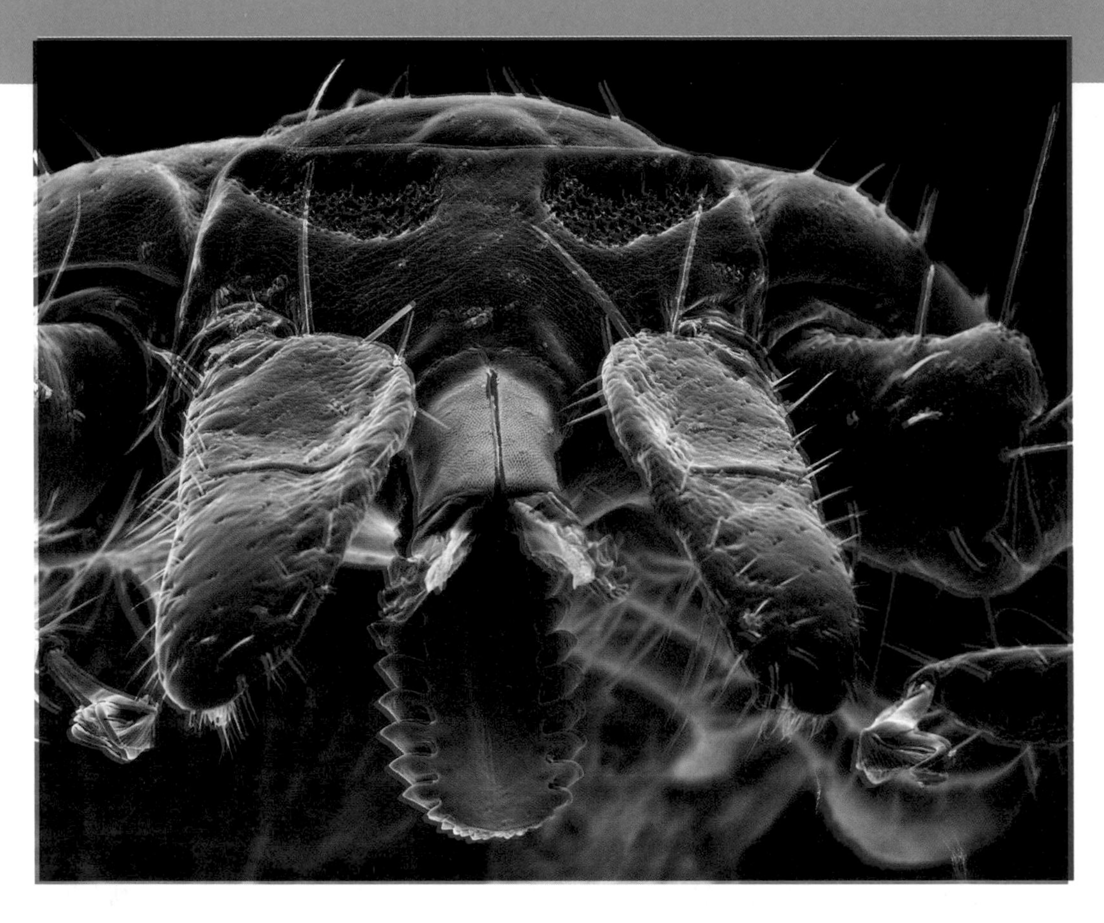

Ticks drink the blood of animals like deer, mice, dogs, and humans. Special mouthparts make a hole in the host's skin. The host doesn't feel it.

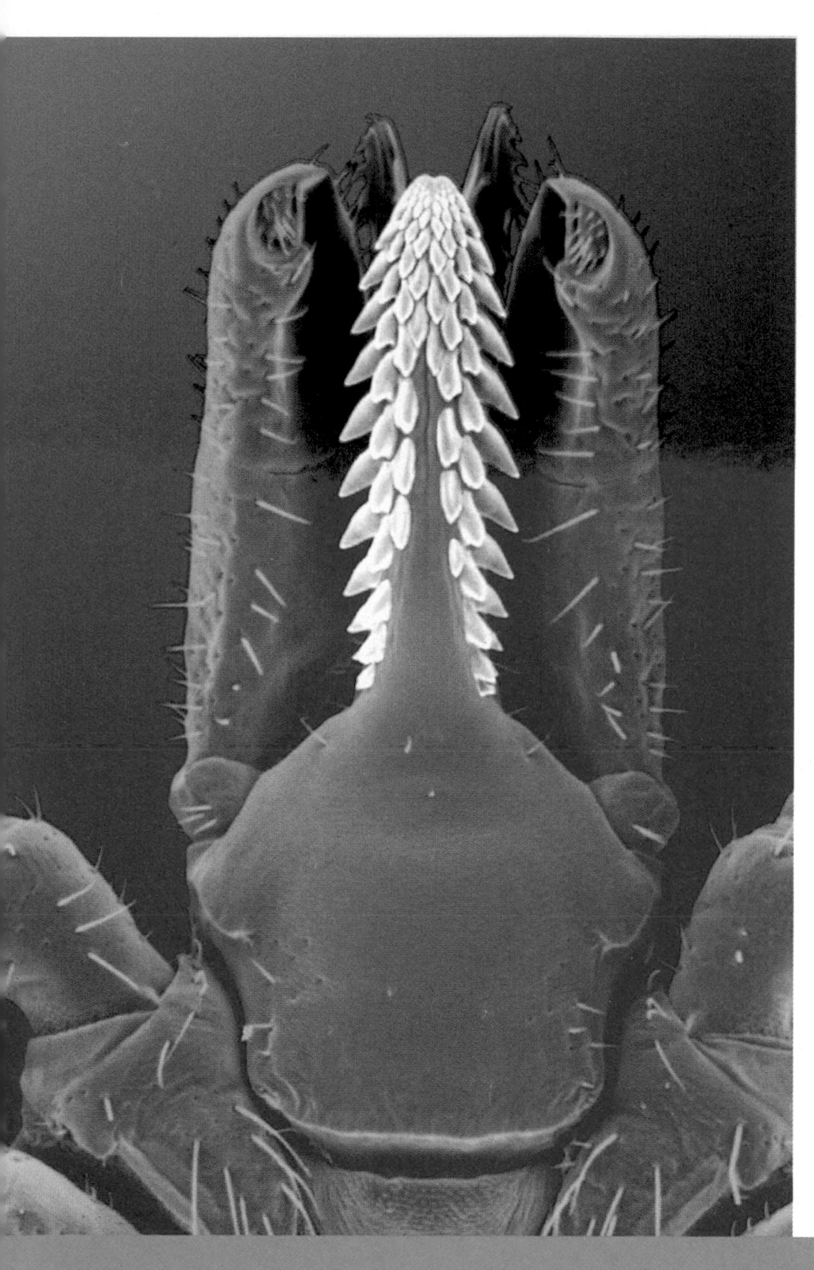

Then, ticks push their **hypostome** into the hole. Ticks suck blood through the hypostome. It has sharp points that face backwards. This makes the tick hard to pull out.

Which Animals Attack Ticks?

These birds are called oxpeckers. They eat ticks and other **parasites** from rhinoceroses' skin. Birds, spiders, and wasps are tick **predators.**

People do not like fire ants because it hurts when they bite. But fire ants eat ticks. And that's good.

Where Do Ticks Live?

Ticks live all over the world. Most live on the ground. They live in grassy meadows, **deciduous forests, coniferous forests,** and weedy fields.

When ticks are not on the ground, they are on a **host.** Deer ticks can give **Lyme disease** to people who are their hosts.

How Do Ticks Move?

Ticks use their legs to crawl on the ground or on a blade of grass. They hold on to **hosts** with their two front legs. Ticks cannot fly or jump.

Ticks have claws at the ends of their legs. Claws help them to hang onto things. This is helpful because their hosts are always moving.

How Long Do Ticks Live?

The ticks that carry **Lyme disease** live for two years. Other ticks can live for three years. **Male** ticks die after **mating. Female** ticks die after they lay their eggs.

Ticks can live a long time without eating. This is important. It may take a long time to find a **host.** Some ticks wait more than a year for a host.

What Do Ticks Do?

Deer ticks carry **Lyme disease.** Lyme disease is caused by **bacteria.** Ticks pass the bacteria to the **host** when they drink their blood.

Lyme disease starts with a bad rash, fever, and body aches. Wear pants and a long-sleeve shirt if you walk where ticks live. Have an **adult** check your body for ticks when you come indoors.

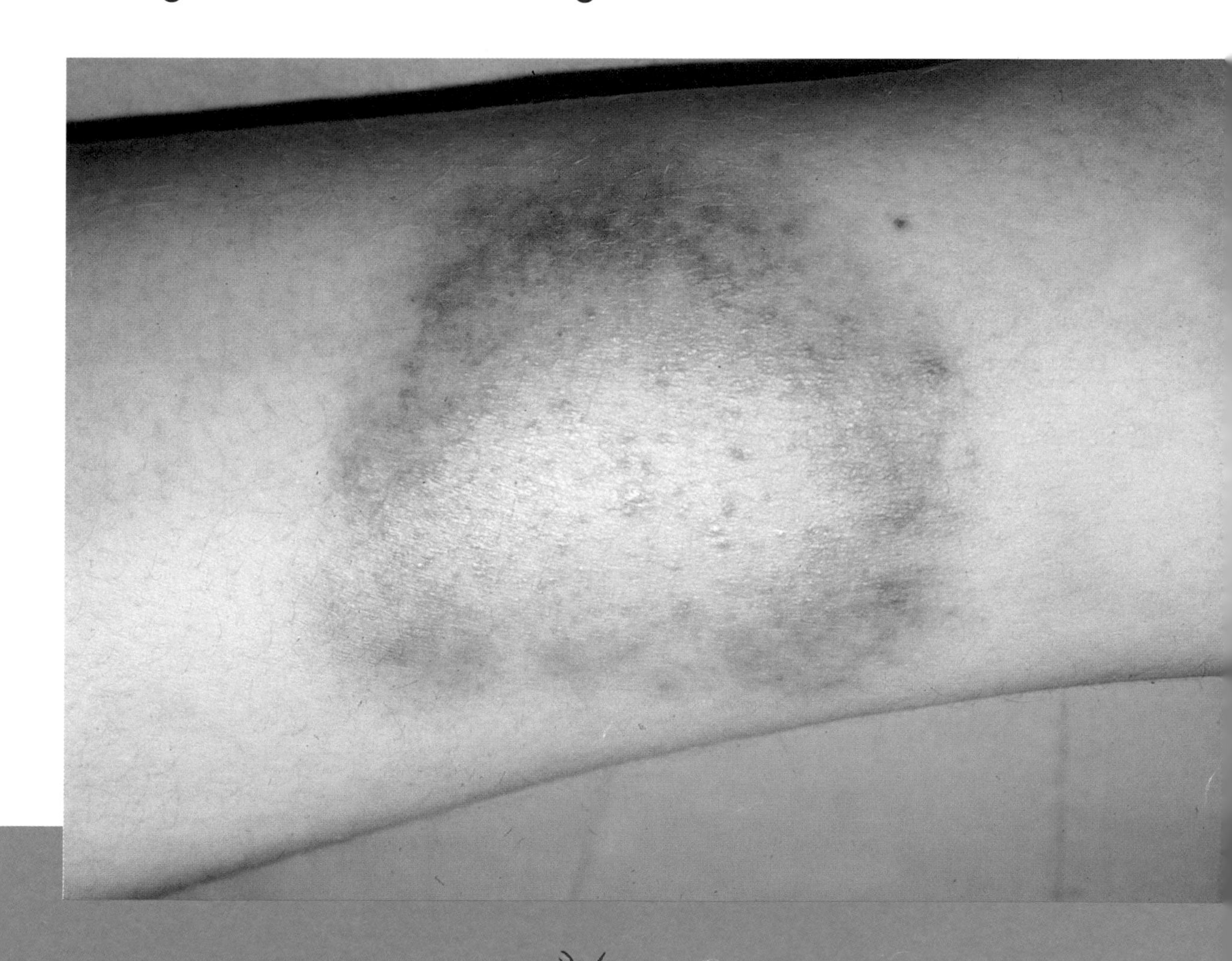

How Are Ticks Special?

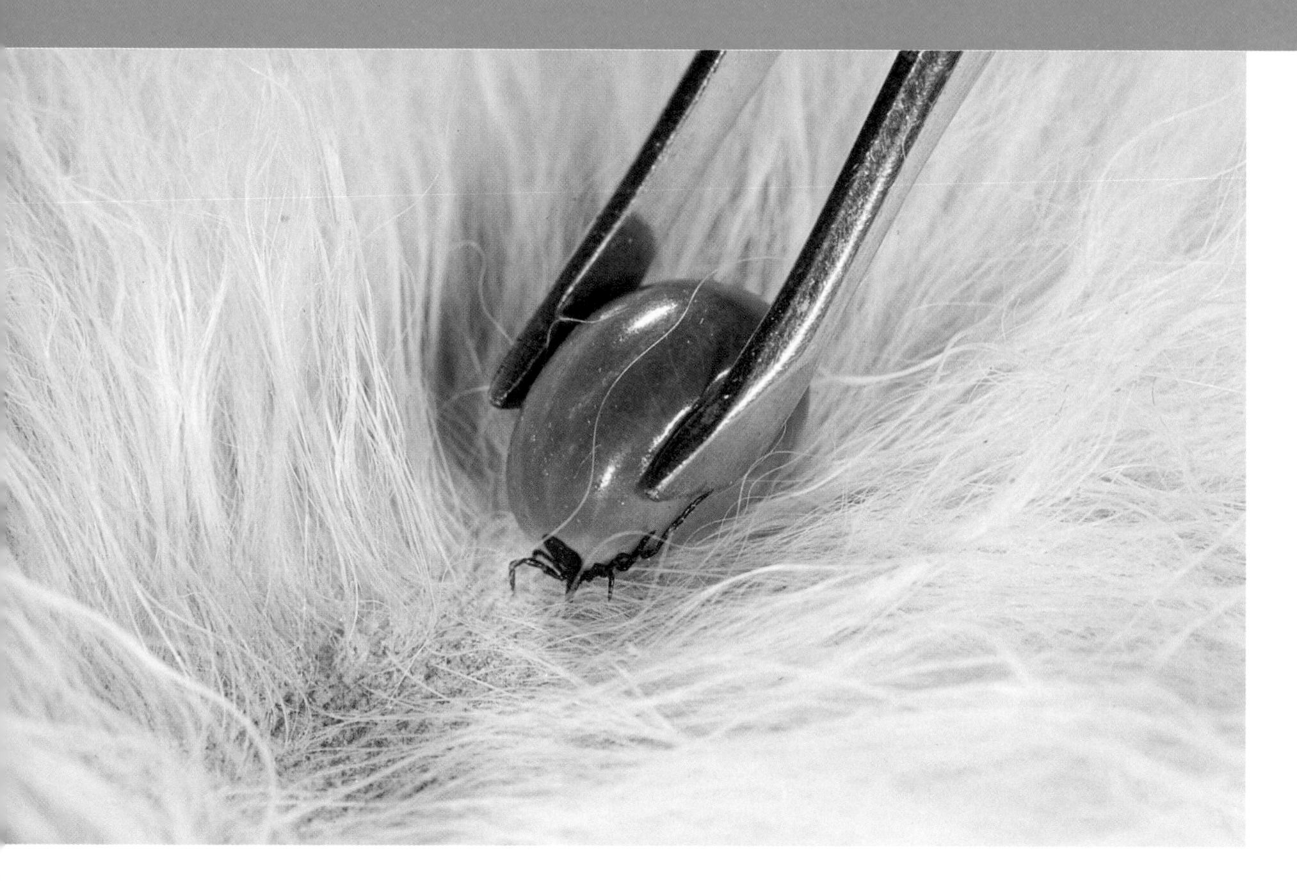

Ticks make a sticky liquid when they bite their **hosts.** This liquid holds the tick's **hypostome** in place so it can eat. It also makes it hard to pull the tick off.

Ticks also make a special liquid in their **saliva.** This liquid stops the host's blood from **clotting.** The tick gets all the blood it needs from just one bite.

Thinking about Ticks

Larvae have six legs and **adult** ticks have eight legs. Which of these ticks is an adult?

The tick's **hypostome** has sharp hooks that face backwards. How do the hooks help the tick?

Tick Map

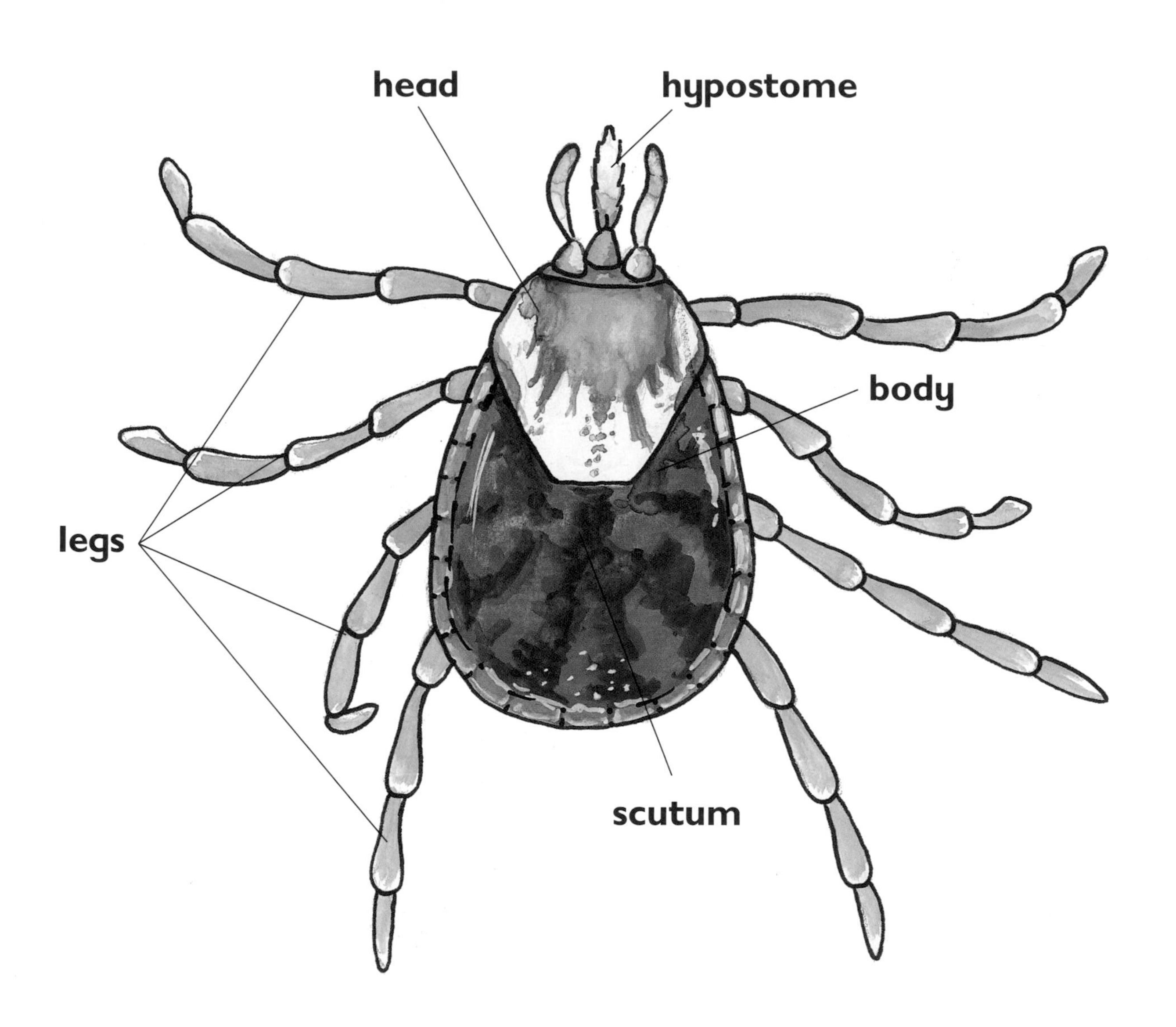

Glossary

adult grown-up
arachnids group of animals that includes spiders, ticks, and scorpions
bacteria (one is a **bacterium**) small living thing that can cause sickness
coniferous forest forest made of trees that never lose their leaves
clotting when liquid blood changes to a solid, like a scab
deciduous forest forest made of trees that lose their leaves in the winter
female girl
hatch to come out of an egg
host animal that a parasite lives on and feeds from
hypostome special mouthpart for getting blood. It works like a straw.
larvae young, six-legged tick
Lyme disease sickness caused by bacteria and carried by ticks
male boy
mate when a male and a female come together to make babies
molt get rid of skin that is too small
mouthparts body parts that help an animal eat
nymph young tick with eight legs
organs body parts with a special job
parasite animal that lives on or in another animal and feeds from that animal
predator animal that hunts other animals for food
questing when ticks search for a host
saliva watery liquid that comes from the mouth
scutum hard plate covering the back of a hard tick

More Books to Read

Merrick, Patrick. *Ticks.* Channassen, Minn.: Child's World, 1997.

Weitzman, Elizabeth. *Let's Talk About Having Lyme Disease.* New York: PowerKids Press, 1998.

Index